FARMER LLAMA'S FARM MACHINES

BALERS

BY KIRSTY HOLMES

WITHDRAWN

BEARPORT
PUBLISHING

Minneapolis, Minnesota

Library of Congress Cataloging-in-Publication Data is available at www.loc.gov or upon request from the publisher.

ISBN: 978-1-64747-542-0 (hardcover)
ISBN: 978-1-64747-549-9 (paperback)
ISBN: 978-1-64747-556-7 (ebook)

For more information, write to Bearport Publishing, 5357 Penn Avenue South, Minneapolis, MN 55419. Printed in the United States of America.

IMAGE CREDITS

All images are courtesy of Shutterstock.com, unless otherwise specified. With thanks to Getty Images, Thinkstock Photo, and iStockphoto. Cover - NotionPic, Tartila, A-R-T, logika600, BiterBig, robuart. Aggie - NotionPic, Tartila. Grid - BiterBig. Farm - Faber14. 2 - studioworkstock. 5 - Mascha Tace. 6&7 - studioworkstock, Jurgis Mankauskas. 8 - Nsit, Peyker. 9 - Ekaterina_Mikhaylova, Mutantur. 10&11 - studioworkstock. 12&13 - studioworkstock, N. Mitchell. 14&15 - studioworkstock, MicroOne, Borodatch, LUMIKK555, allstars, Orest Iyzhechka. 16&17 - Lidiia, studioworkstock. 18 - studioworkstock. 20 - Mascha Tace. 21 - DRogatnev. 22 - SKARIDA, Iconic Bestiary, Dudarev Mikhail. 23 - StockSmartStart, ann131313. 24 - photomaster.

CONTENTS

DOWN ON THE FARM!

Welcome to Happy Valley Farm. You must be the new **farmhand**. My name is Aggie, and I'm a farmer llama.

Hay there! Nice to meet you!

This is one of the busiest times of the year on the farm. Let's get you trained in!

What You Need to Know

Round or square? ☐

How does a baler move? ☐

How does grass become a bale? ☐

How do you say *scythe*? ☐

WHAT IS A BALER?

BALER

STRAW

TRACTOR

A baler is a machine that makes bales. Balers are attached to tractors, which pull them along and give them power.

Hay is dried grass. It is used to feed **livestock** on the farm. Straw comes from what is left of wheat and other crops. It is used for animal bedding. Both hay and straw are made into bales.

BALE

BALER

HAY

Munch, munch...
I'll be with you in
a minute!

BEFORE BALERS

Before farmers had machines, they had to make hay and straw bales by hand. Farmers would cut the plants with a scythe and then leave them in the sunshine to dry.

Scythe is a funny word! You say it siTHe.

SCYTHE

THE ONLY WAY DOWN IS TO SLIDE OFF!

Then, farmers would start piles of hay and straw. Someone would stamp on the piles along the way as more straw and hay was added. As the piles grew, the person on top would go higher and higher!

ROUND . . .

Bales can be round or square.

ROUND BALES ARE . . .

LARGER.

QUICKER TO MAKE.

I'll just check this hay to make sure it's good to eat . . . . *nom, nom . . .*

GOOD FOR FARMS WITH LOTS OF LIVESTOCK.

OR SQUARE?

SQUARE BALES ARE . . .

SMALLER.

EASIER TO MOVE.

GOOD FOR SMALLER FARMS AND **STABLES**.

PARTS OF A BALER

Each part of a baler does a job.

TRACTOR
The tractor powers and moves the baler.

STUFFER
The stuffer pushes the pile into shape.

Let's take a look at a baler.

TAILGATE

The tailgate opens, and the bale is dropped to the ground.

SQUARE BALERS FORM SQUARE BALES.

BALE CHAMBER

The bale forms in the **chamber**.

PICKUP

The plant is collected at the pickup and is scooped into the machine.

HAY! WHAT'S THAT?

Welcome to my favorite lesson. It's time for hay tasting! This hay is abso-llama-lutely tasty! These are the three most common types of hay.

TIMOTHY

- Also called herd's grass or meadow cat's tail
- Found in Europe
- Grows to around 3 feet (1 m) tall

CLOVER

- Good for sheep, horses, and small animals
- Sometimes mixed with other grasses
- Can become **moldy** when baled

TIMOTHY

CLOVER

HAY TASTING

ALFALFA

- Part of the pea family of plants
- Grown all over the world
- Eaten by cows and lots of other animals

Say this tasty treat like al-FAL-fuh.

ALFALFA

FACT FILE

Let's find out more about those big bales!

Round bales can weigh 2,000 pounds (900 kg) or more!

If a bale has too much water in it, it can burst into flames!

RECORD BREAKERS

LARGEST PYRAMID

The largest **pyramid** of bales ever built was made up of 1,500 bales of hay. It was almost 95 ft (29 m) tall!

It was more than 80 ft (24 m) wide at the bottom.

LARGEST MAZE

In 2011, the largest bale maze in the world was built in Idaho.
It was made using 3,202 bales and had a path that was 1.6 miles (2.6 km) long!

Yeehaw! That's a lot of bales!

GET YOUR LLAMA-DIPLOMA

Awesome job! You finished your training. Time to see what you've learned. Get all the questions right and you'll get your llama-diploma!

Questions

1. What pulls a baler along?

2. What is cut down and dried to make hay?

3. Are square or round bales larger?

4. What comes out of a baler's tailgate?

5. How do you say *scythe*?

You made that look easy! Welcome to the Happy Valley Farm family!

CERTIFICATE
OF APPRECIATION

LLAMA-DIPLOMA

HAPPY VALLEY FARM

Farmer Llama

Download your llama-diploma!

1. Go to **www.factsurfer.com**

2. Enter "**Balers**" into the search box.

3. Click on the cover of this book to see the available download.

ON YOUR BIKE

The hard work has all been done by the baler, so we have a little time left over at the end of the day. It's time to play . . .

STEP ONE

Get some round bales

STEP TWO

Find a hill

STEP THREE

Get your bike

GLOSSARY

CHAMBER a closed-in space

FARMHAND a person who works on a farm

LIVESTOCK animals that are kept for farming

MOLDY covered in mold, which makes it break down and rot

POISONOUS dangerous or deadly when eaten

PYRAMID a shape with triangular sides that meet at a single point

STABLES buildings where horses or cows are kept

INDEX